Yassine Izarif

Geological exploration of a virgin area (Kerkouz)

Yassine Izarif

Geological exploration of a virgin area (Kerkouz)

in the Jbilets of Marrakech. Mapping, borehole study, tectonics, geochemistry and magnetic susceptibility

ScienciaScripts

Imprint
Any brand names and product names mentioned in this book are subject to trademark, brand or patent protection and are trademarks or registered trademarks of their respective holders. The use of brand names, product names, common names, trade names, product descriptions etc. even without a particular marking in this work is in no way to be construed to mean that such names may be regarded as unrestricted in respect of trademark and brand protection legislation and could thus be used by anyone.

Cover image: www.ingimage.com

This book is a translation from the original published under ISBN 978-620-2-54596-9.

Publisher:
Sciencia Scripts
is a trademark of
International Book Market Service Ltd., member of OmniScriptum Publishing Group
17 Meldrum Street, Beau Bassin 71504, Mauritius
Printed at: see last page
ISBN: 978-620-3-48227-0

Contents :

<u>**Acknowledgements**</u>

At the end of this modest work, we would like to thank our supervisors Mrs El Yaziji Assia and Mr Rednaoui Ahmed who spared no effort to support us throughout the period of our training course.

We also express our gratitude and sincere appreciation to all my teachers at the Department of Applied Geology of the Institute of Mining in Marrakech and especially to Mr. Nadeif and Mr. Zehouani for all their valuable explanations. Our sincere thanks also go to all the people in charge of the exploration office of the Hajjar mine: The head of exploration Mr.Ouadjou Abd Elmalek, Mr.Zaghari Mustapha and Mr.Oulaarif Mohammed, we forget the team of Techsub in charge of the drilling works in our study area who helped us in the realization of this work and who did not cease to put at our disposal all the means necessary to the achievement of our work.

Our deepest thanks and appreciation go to our family for their moral and financial support during this internship and throughout our graduate studies. Finally, we would like to thank all those who have contributed in any way to the elaboration of this report.

<u>**Signing session**</u>

May this work bear witness to my respects:

<u>**To our parents:**</u>

Thanks to their tender encouragement and sacrifices, they were able to create the loving and conducive climate for our studies.

No dedication could express our respect, consideration and deep feelings towards them

We pray to God to bless them, to watch over them, and to make them proud of us.

To our sisters and brothers.

<u>**To all our teachers:**</u>

Their generosity and support obliges us to pay them our deepest respects and our loyal consideration.

<u>**To all our friends and colleagues:**</u>

They will find here the testimony of an infinite loyalty and friendship

<u>**Part One: General**</u>

I.   Structural areas of Morocco :

Morocco is known as a geologist's paradise because of the diversity of its geological formations and the quality of the outcrops. It is located at the western end of North Africa and is the most eroded part of the continent.

In the course of its geological history, Morocco has experienced several orogenic cycles (Precambrian, Hercynian and Alpine cycles). These cycles are at the origin of the current structural configuration of the country in five major structural domains (Piqué and Michard, 1989) (Figure 1).

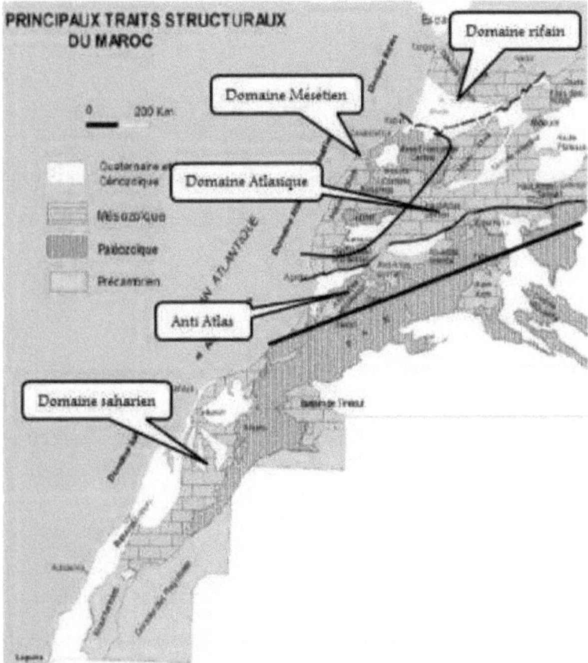

<u>**Figure 1: The main structural features of Morocco.**</u>

- **Rifa domain :**

Represents the northern limit of Morocco. It is made up of allochthonous nappes linked to the alpine orogeny overlapping the Mesetian domain. The Paleozoic terrains are known in the internal zones of the chain (Paleozoic nappes or Gomarides).

- **Mesotian domain:**

It represents the Moroccan Hercynian domain and is subdivided by the Middle Atlas into the Western Meseta represented by the Central Massifs of Morocco, the Rehamna and the Jbilets, and the Eastern Meseta characterised by several buttonholes, including the Midelt buttonhole (Mibladen- Ahouli)

To the SW and Mekam, Delolom to the NE). It consists of a Paleozoic basement covered unconformably by tabular Meso-Cenozoic series (A. Michard, 1976).

- **Atlas area :**

Extended across the Meseta (Middle Atlas) and between the Meseta and the Anti-Atlasic domain (High Atlas). The Permo-Mesozoic and Cenozoic terrains are structured during the Atlas orogeny (Upper Jurassic and Tertiary folding phases).

- **Anti-atlasic and Saharan domain :**

> **The Saharan Domain :**

Constitutes part of the West African craton (Fabre, 1971). This domain is made up of terrains of Lower Proterozoic age, intensely deformed and metamorphosed during the Eburnian orogeny (2000 Ma). The whole is covered in the north by the undeformed Palaeozoic rocks of the Tindouf Basin.

> **The Anti-Atlasic Domain**

Consists of a basement structured by the Pan-African orogeny (680 and 570 Ma) (Leblanc and Lancelot, 1980). Its terminal Proterozoic and Palaeozoic cover is affected by relatively moderate Hercynian deformation. The Mesozoic and Cenozoic cover is undeformed and of low strength.

II. The Jbilets massif:

1. The geographical and geological setting of the Jbilets :

The Jbilets (small mountains), are a set of hills and rocky plains of Paleozoic, folded and metamorphosed terrain oriented in an E-W atlasic direction over a length of about 170 km and 7 to 40 km wide (Huvelin, 1977).

The Jbilets Massif is characterised by the intensity of the pre- to syn-orogenic magmatic activity that this region experienced towards the end of the Carboniferous (Huvelin, 1977).

The Hercynian islands of the Jbilets plunge under the Bahira plain to the north and the Haouz plain to the south, of Miopliocene and Quaternary age. To the west, they are limited by the Jurassic-Cretaceous hills of Mouissat and to the east, by the Middle Atlas mountains of Beni-mellal.

2. Structural subdivisions of the Jbilets :

The Jbilets Massif is subdivided into three major units with distinct lithostratigraphic and tectono-metamorphic characteristics (Figure 2).

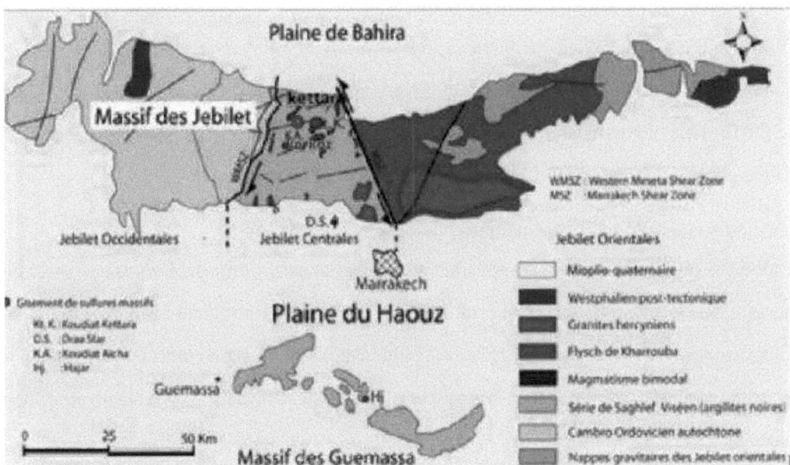

Figure 2: Structural zonation of the Jbilets massif (Maier et al. 1988).

The Western Jbilets :

Extends from Kettara (on the road to Safi) to Jbel Irhoud. It is composed of Cambro-Ordovician schistogresous terrain, Westphalo-Permian conglomerates and Permotriassic sediments (sandstones and pelites) and basalts. This unit corresponds to the stable block or "coastal mole". It is little or not metamorphosed and affected by submeridian folds of hectometric to kilometric amplitude.

The unity of the central Jbilets :

This is the unit that encases the Draa Sfar deposit and extends from the Kettara mine to the Sidi Bou Othmane (Marrakech-Casablanca road); it is subdivided into two large series. The first series is of Superior Visée-Namurian age (Carboniferous), it is composed of the Flyschs of Kharrouba, it is an alternation of sandstones and pelites, which became quartzites and shales by the effect of the granodioritic intrusion, Overlain by the Sarhlef shales (volcano-sedimentary formations where the limestones at the top or intruded have a lenticular shape) and intruded by acid and basic massifs (granite, granodiorite, gabbros and volcanic tuffs); This series is of subsident type.

◆ **Oriental jbilets:**

Spreads out from Sidi Bou Othmane towards the East (main Casa-Marrakech road), it is a unit formed by nappes of charriages (from the Ordovician to the Devonian) which come from the East to the West by an elevation during which the cover often detaches at the level of the Silurian plastic pelites, and covers in pseudo concordance sometimes the top of the Flyschs of Kharrouba, sometimes the base of the Sarhlef Schists (HUVELIN 1977). These nappes are gravity-type slides that are called olistostromes.

3. **Stratigraphy and sedimentology :**

Despite the effect of tectonics responsible for the structuring of the Jebilets massif with the presence of metamorphism and the setting up of magmatic intrusions responsible for the mineralisation, the stratigraphic evolution could be defined with two units corresponding to the upper ante-Visian and upper Visian periods and showing the following stratigraphic succession:

i. **The early Ante-Visian period :**

   Represented by the following floors, namely

   v <u>Cambrian</u>: represented by detrital formations with episodic volcanic facies. It is represented mainly in the western Jbilet.

   v <u>Ordovician</u>: present in the western Jbilets. Its lithostratigraphy is essentially represented by clayey pelites, detrital sediments and conglomeratic sandstones.

   ■*J* <u>Silurian</u>: known only from the eastern Jbilets. It is composed of graptolite shales and phtanites.

   ■*J* <u>Devonian</u>: formed by red conglomerates, massive limestones and quartzite sandstones in the Skhrats countries. At the level of the eastern Jbilets, it consists of bivalve shales, shales with limestone banks and alternating sandstones and shales.

   ■*J* The <u>Tournaisian</u> and <u>Lower Viséan</u> are absent.

ii. **The Upper Viséan period :**

   Corresponds to a major transgression, originating from the East or NE (Beauchamp,

1984; Piqué, 1994). It comprises two lithological units, the Kharrouba flysch in the eastern half of the Jebilets and the Sarhlef shales in the central Jebilets. The structural relationships are still debatable: according to Huvelin (1977), these two sets reflect a vertical evolution, with the Kharrouba series at the base, whereas according to Gaillet (1979) and Bordonaro (1983), they correspond to a lateral variation in facies.

4. The deformation phases of the Jebilets :

Deformation phases	Consequences
Cambrian-Ordovician subsidence movements	Very thick detrital sediments (7600m for the Cambrian alone).
Lower Devonian phase	Slumping, causing an unconformity of Pragian shales and limestones on the Graptolith shales. Folding, emersion and unconformity of the Lower Devonian on the older terrains.
Later Middle Devonian and earlier Upper Visean phase	Folding, if not scaling, resulting in an unconformity of Upper Viséan bioclastic limestone on the Lower or Middle Devonian.
Upper Viséan phase	The synsedimentary scaling of the ante-Viséan terrain will favour the setting up of the nappes: The ollistroms.
Post-Visean phase	Folding and nappe emplacement (Eastern Jbilets) Schistosity, metamorphism and granitisation (Central Jbilets) .
Post-granitisation compression phase	Folds with Kink-band with crenulation schistosity.
Post-granitisation distension phase	Establishment of microdiorite vein (central jbilets).
Late Hercynian compression phase	Charriages and late Hercynian decays.

Block tectonics	
	Hercynian glyptogenesis deposition of the Westphalo-Permian in major unconformity on the post-Visean basement. Installation of a network of subvertical faults.
Intra-Permian folding phase	
	Folding prior to the deposition of the Permo-Triassic. Rejeux and possible extension of the subvertical fault network.
Distension phase after the intra-permian folds	Establishment of the gabbro vein (western jbilets) and the dolerite veins.

<u>**Table 1: The deformation phases of Jbilets.**</u>

5. Magmatism :

According to Huvelin (1977), the end of the Carboniferous period in the Jbilets Massif was marked by intense pre-orogenic magmatic activity, but this was later than the Middle Upper Viséen and occurred in a distensive geodynamic context. The basic and acidic magmatic rocks derived from the same parent magma (AARAB, 1984). Magmatic activity in the Jbilets occurred mainly in the central and eastern Jbilets. It can be divided into three main episodes or magmatic activities that were timed in relation to the paroxysm of the Hercynian deformation (HUVELIN, 1977).

The first activity, known as pre-tectonic, is composed of a magmatic procession dominated by basic rocks (gabbros and dolerites) and ultrabasic rocks (wherlites) associated with acidic terms (plagiogranites) of an abyssal tholeiitic nature. It is contemporary with the opening of the Visean basin, prior to Hercynian tectonics.

The second activity, described as orogenic, is marked by the presence of syn- to late-tectonic calc-alkaline granites, emplaced in the form of circumscribed batholiths. These granites, contemporary with the post-Vetian crustal shortening (Lagarde and CHOUKROUNE, 1982), have developed a large halo of contact metamorphism.

The third actvity, known as post-orogenic, corresponds to the development of a cluster of sub-meridian microdiorite veins that cut through the orogenic granites. These veins are lamprophyres of kerstites type of Triassic age. They contain enclaves of various types which testify to the existence of a deep basement (gneiss, migmatites, granites) of Proterozoic age. According to HUVELIN (1977), the emplacement of these microdiorites is linked to a relaxation phase subsequent to the major folds linked to the Hercynian orogeny. This magmatic episode is probably linked to the opening of the Atlantic Ocean.

6. Metamorphism :

The metamorphic rocks that dominate the area are the result of regional metamorphism (Sarhlef schists) and contact metamorphism (andalusite spotted schists, and/or cordierite at the edges of granitic intrusions), and sometimes both types combined. The region of Bramram, Tabouchent and Bamega show a contact metamorphism developed around the Hercynian granites (bearing the same names).

7. The sulphide clusters of the central Jbilets :

- Location of sulphide clusters in the central jbilets :

The central unit of the Jebilets contains several sulphide clusters. With its well-developed iron cap and location on the Marrakech-Safi road, the Kettara deposit was the first sulphide heap discovered in the Jebilets and was exploited from 1964 to 1982 for pyrrhotite sulphur (Essaifi, 2012). Other sulphide clusters have been located at Ben Slimane, Kerkoz, Draa Sfar, Bouhane, Lachach and Kt. Aicha. On the map scale (Figure 3), the sulphide clusters of the Jebilets and their iron caps constitute N-S to NE-SW trending subvertical lineaments (Bernard et al, 1988), parallel to the Hercynian structures (Essaifi and Hibti, 2008).

- The Bouhane, Lachach and Koudiat Aicha deposits and iron caps, offset by the dexter Mesret fault, form the western lineament. These deposits are rich in Zinc and Lead (Essaifi and Hibti; 2008). The Koudiat Aicha deposit contains 3.6 Mt of metal grading 3% Zn, 1% Pb and 0.6% Cu (Lotfi et al. 2006).

- The deposits and iron caps of Kettara, Benslimane and Kerkoz form the central lineament. The Kettara deposit and its southern extension, the Benslimane deposit, are poor in Zn and Pb (0.6% Cu at Kettara and 0.8% Cu at Benslimane; Huvelin, 1977).

- The eastern lineament includes the Draa Sfar deposit in the south and the Nzalet el Harmel and Ben El Garn iron caps in the north. The Draa Sfar deposit contains 10 Mt of metal grading 5.3% Zn, 2% Pb and 0.3% Cu (Marcoux et al., 2006).

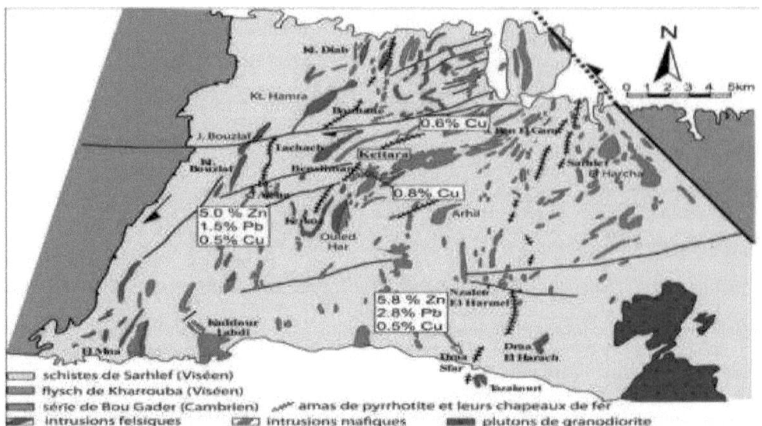

Figure 3: Location of the main pyrrhotite sulphide clusters represented by their iron caps (in Essaifi 2012; modified from Huvelin, 1972). Base metal contents of some sulphide clusters are indicated (after Huvelin, 1977; Lotfi et al, 2008; Marcoux et al, 2008). (Huvelin, 1977).

- The geodynamic context :

Volcanogenic sulphide clusters (VMS) are the manifestation of oceanic hydrothermal circulation. Their formation requires the following elements (Figure 4):

A heat source to provide energy for the system;

A reaction zone, where high temperature acidic fluids react with rocks. These reactions take place at around 400°C and their balance can be written with reactions such as Basalt+water=epidote+ actinote+ quartz+ ions in solution. They therefore form a fluid rich in ions, in particular Fe and geochemically related elements such as Cu, Zn and Pb;

> An impermeable roof and a discharge zone, both of which channel fluids along the fractures;

> A precipitation zone on the ocean floor where warm, metal-laden water comes into contact with cold water to build a sulphide mound;

> Percolation of surface water replacing hot water. Economically viable deposits of large size are made by fluid circulation.

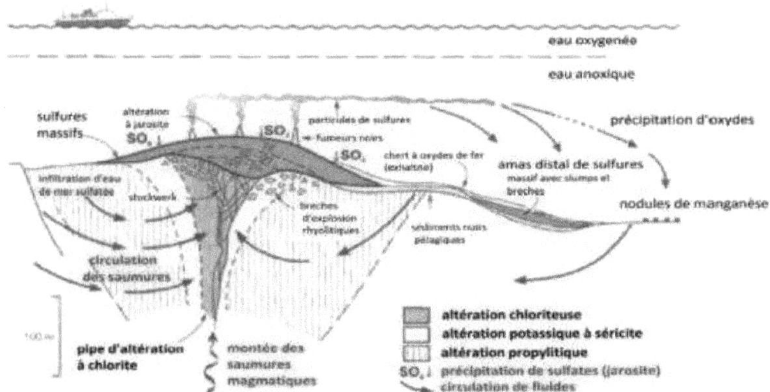

Figure 4: The general principle of setting up VMS mineralisation in the bottom.

- **Alteration related to sulphur clusters :**

The hydrothermal systems of the VMS are linked to two types of alteration depending on the depth. In the deep domain, there is a depletion of potassium, magnesium and iron. Near the surface, there is a clear zonation between sericitisation (the loss of Na, Mg, Si and Ca in the least affected areas) and chloritisation (the areas most altered by the loss of the above elements). Silicification is the result of the replacement and precipitation of dissolved silica further away. This zonation is a good field indicator of the proximity of the deposit. (Gibson and Galley, 2007), (Figure 5).

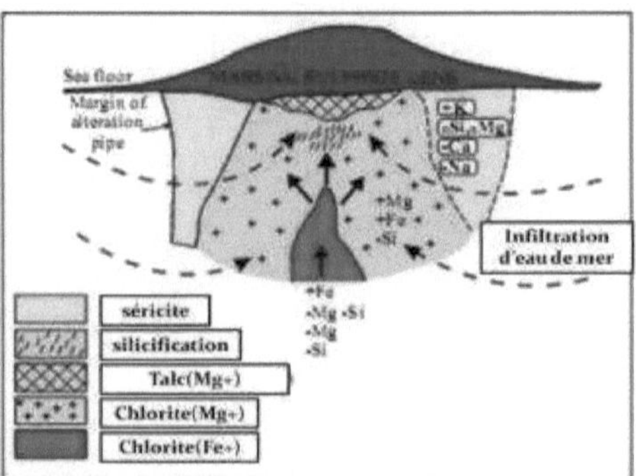

Figure 5: Schematic cross-section of the weathering stack associated with a VMS (Gibson and Galley, 2007) shows the layout of the main weathering zones, and the ion movements.

- Economic interest :

These clusters consist mainly of pyrrhotite (75-95%) with sphalerite, galena, chalcopyrite, arsenopyrite, pyrite, marcasite and magnetite. Other minerals present in trace amounts include stannite, cassiterite, cobaltite, native bismuth, bismuthinite and native silver. Geochemical investigations of the main sulphides in the deposits show that the concentration of major and trace elements in the main phases (pyrrhotite, sphalerite, galena, arsenopyrite and chalcopyrite) is variable from one area to another.

Part Two: Contribution to the geological study of Kerkoz (Central Jbilets)

I. Introduction :

1. Subject :

The present study is part of our training as technicians specialised in applied geology. It concerns the Kerkoz area, which has become a study and exploration sector of the CMG (Compagnie Minière des Guemassa).

Our study is focused on the exploration of the Kerkoz zone as a virgin zone and more precisely the iron caps and their continuities on this zone, the distribution of chemical elements to better delineate the cementing zone, then the structural and metallogenic characterization of the prospect existing in this zone.

Finally, an attempt will be made to characterise the hydrothermal alteration and deformation associated with mineralisation and the correlation between the base metals.

Therefore, this study aims at the geological characterisation (lithostratigraphy; petrography; geochemistry and structure) of iron caps.

2. Methodology :

In order to achieve the targeted objectives, several studies were carried out.

K Map of the Kerkoz area at a scale of 1:2000 ;

K Exploit the exploration work, i.e. scraping and core drilling, in order to carry out a mineralogical and petrographic description of the facies encountered;

K Tectonic study of the area to locate outcrops of mineralized lenses (Iron Cap);

K Collect samples from the mineralized areas for chemical analysis of any base and precious metals present;

K Make magnetic susceptibility curves across 3 boreholes to deduce the existence of magnetic minerals in the iron cap zones (oxidation zone, transition zone and cementation zone) using the magnetometer.

3. Geographical location of Kerkoz (central jbilets) :

Our study area is located in the central jbilets and more precisely it concerns the iron caps outcropping in the healthy sarhlef shales, considered as a very promising sector by the presence of these iron caps rich in mineralization.

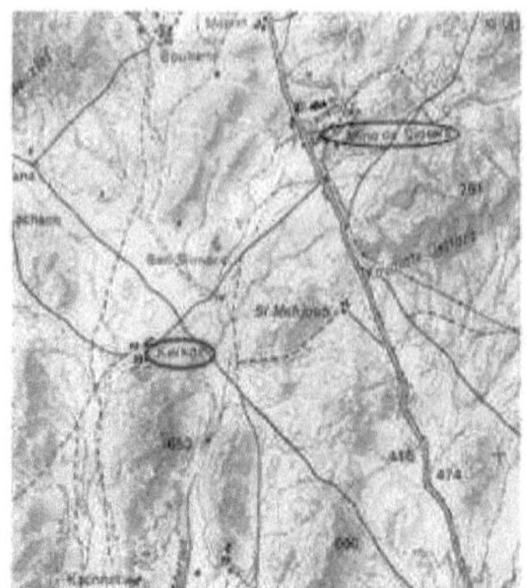

<u>Figure 6: Location of Kerkoz</u>

The Kerkoz sector is located 37 km north of Marrakech and 120 km from Safi, this sector is limited to the north by Douar Ben Slimane, to the west by Douar Cheikh Ali, to the east by Douar Si Mahjoub and to the south by TechNet. It is mainly interesting for the following mineralizations: Ochre, Copper, and Iron Oxides.

II. Mapping of the Kerkoz area (central jbilets) :

1. Introduction :

Mapping is one of the most important methods of mineral exploration, based on the search for metallotects of mineralisation and the representation of different facies on surface, which is why mining companies need maps for a global approach during the general prospecting phase in order to understand the geological and metallogenic context of metallic showings.

In order to determine the different geological formations outcropping on the surface of the Kerkoz study area, to locate the different phenomena and structures that can give information on mineralization (iron caps) as well as to follow the distribution of this mineralization in our terrain, it was recommended to carry out a detailed mapping of this area, with more developed methods using a software that is well known in the area by its performance and credibility "Mapinfo".

2. Method of making the geological map :

In order to establish a geological map of the Kerkoz area, we based our work on measurements of the X, Y and Z coordinates of the various terrain features using a GPS set for a Merchich projection (such as schistosity, faults, iron caps, dykes, seams, drainage network, tracks, and overburden), as well as the measurement of the various directions and dips of these features (using a compass).

After the practical stage in the field, we used the software "MapInfo" to process all the coordinates we took in the field according to well-targeted Excel tables (such as schistosity, faults, hydrographic network, lithostratigraphy, tracks, coverings...). This software presents the different points measured in the field, and it is the role of the geologist to correlate these points and to complete and interpret all that is missing on the geological map.

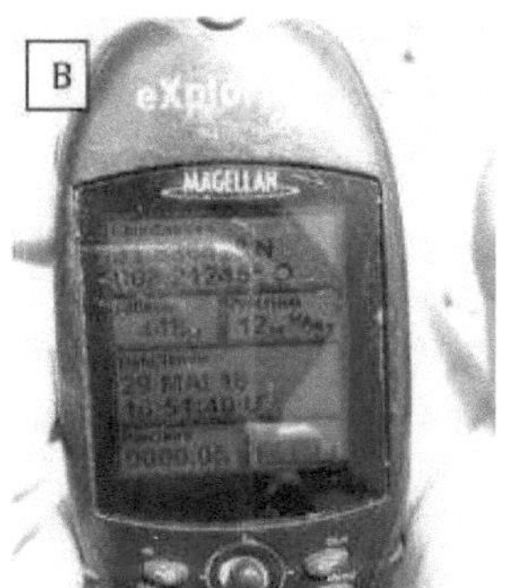

Photo 1: Equipment used: A- Hammer, Compass and B- GPS

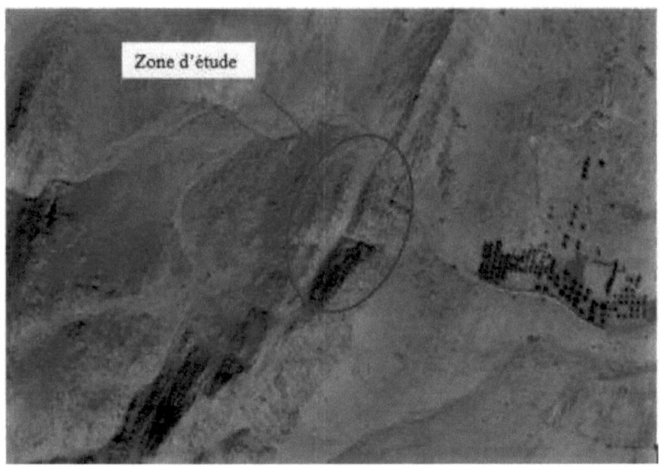

<u>**Photo 2: Satellite photo of the study area**</u>

3. Map made :

In order to properly determine the different geological formations outcropping in the Kerkoz study area and to locate the mineralised structure as well as to follow the iron caps and their continuities, a mapping of the area was recommended.

The coordinates of the different structures and deformations (schistosity, fault) were taken by GPS and then the map was digitised using the MAP-INFO software.

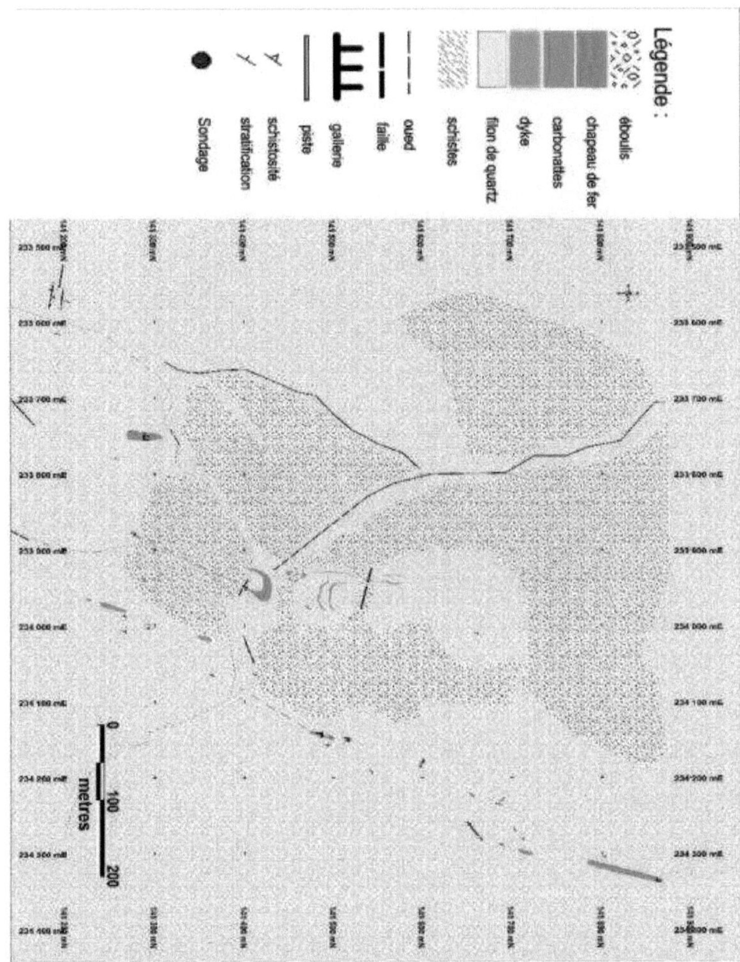

Figure 7: Geological map of the study area (Kerkoz)

The map shows that the area is dominated mainly by the Sarhlef shales, which are composed of alternating pelite and sandstone and which host the iron caps distributed in the area in a way that indicates the mode of deformation of the area. There is also an outcropping magmatic component which manifests itself in dolerite dykes and white quartz veins.

4. **Description of the lithological facies encountered :**

The lithological study of Kerkoz is based on direct observation of the different facies in the field.

Photo 3: The Sarhlef Shale

Or slab schists, which are the most abundant facies in our study area, are a fine-grained facies that is a witness to regional metamorphism.

It is greyish to black in colour and poor in metamorphic minerals. The direction of the schistosity is parallel to that of the iron cap (N30 to N50°).

These shales change colour in places (usually near the iron cap), they become reddish (hematized) or whitish (carbonated) which is due to the circulation of hydrothermal fluids.

Photo 4: Reddish oxidised shale.

Photo 5: The whitish shale

b) Carbonates :

Forming lenses interspersed within the metamorphic schists, the carbonates represent a well-observed layering with respect to the schistosity (S 1) of the domain schists. The carbonates were well distinguished by the bright effervescence on the planes of the fresh breaks and by the calcite observed in these planes.

Photo 6: Carbonates found in the study area

c) The Iron Hats:

In the field, iron caps are referred to as the limonite masses formed by alteration of deposits containing iron sulphides. Then, by extension, we speak of iron caps of metallic sulphide deposits, i.e. the zone where the oxidation of these sulphides takes place. This zone, which varies in strength from 4m to 20m and in direction from N10 to N30.

Photo 7: Photo of an outcrop of the iron cap

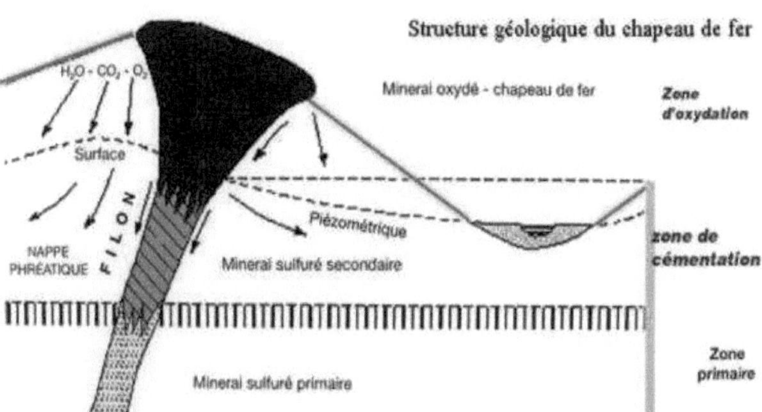

Figure 8: Geological structure of the iron cap

Photo 8: Photo showing the shale-iron cap contact

For the most common minerals, compounds, we have :

- Limonite :

Limonite is the most dominant mineral in the study area.

Photo 9: Limonite in the iron cap.

- Goethite

In the field, goethite is found in a mamelled form with a shiny black colour.

Photo 10: Mammalian goethite

- Hematite :

Hematite occurs on the ground in the form of films or spots in shales near the iron hat.

Photo 11: Hematite with limonite and malachite

Ochre :

Ochre is also present in outcropping iron caps but its colour changes according to the compound present: red for hematite, brown for limonite and yellow for goethite...

- Malachite and Azurite :

Malachite and azurite are often present in the iron caps of our domain as indications of the presence of sulphides at depth.

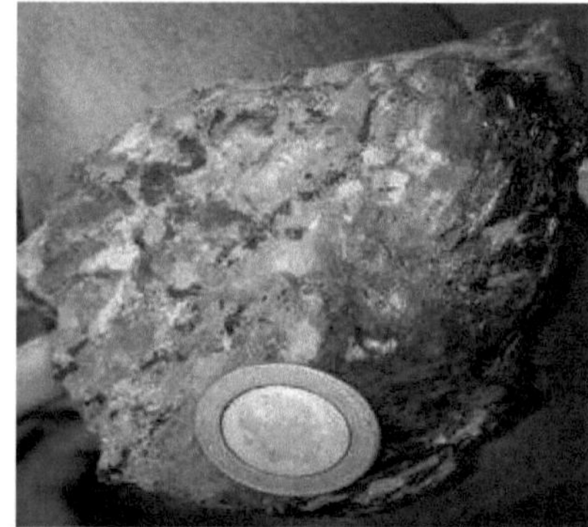

Photo 12: Photo showing malachite with hematite.

Photo 13: Photo of azurite with malachite.

d) Milky quartz :

In the Kerkoz area, the white quartz occurs in the form of hydrothermal veins generally oriented from N10 to N30, parallel to the Sarhlef shales that they encase and sometimes quartz veinlets are found in the shales near the Oued

Photo 14: Fractured hydrothermal quartz

Photo 15: Contact Quartz vein - Iron cap

Photo 16: Quartz vein

Photo 17: Quartz veinlets in shale

e) Dolerite Dyke :

The dolerite occurs in our study area in the form of dykes (basic) with a small diameter outcropping on the surface with a N 20 direction, fractured, diaclassed, and hosted by the Sarhlef shales. These dykes are 2 to 8 m thick and generally have a brownish colour which becomes greenish due to chloritisation.

<u>**Photo 18: Fractured and diaclassed dolerite dike**</u>

5. **Establishment of the lithostratigraphic logs of drill holes KTSC14, KTSC16 and KTSC17:**

In this section, the data from the field surveys KTSC14, KTSC16 and KTSC17 are discussed.

i.   **Positioning of surveys :**

name	X	Y	Z (altitude)	Length(m)	Situation
KTSC13	-8,21089	31,83729	475	25,00	Not studied
KTSC14	-8,21135	31,83753	478	300,00	studied
KTSC16	-8,21392	31,83631	486	110,00	studied
KTSC17	-8,21405	31,83681	489	162,50	studied
KTSC18	-8,21413	31,83593	465	159,50	Not studied
KTSC19	-8,21109	31,83633	483	In progress	Not studied

<u>**Table 2: Positioning of boreholes in the study area.**</u>

■*S* The KTSC14 survey:

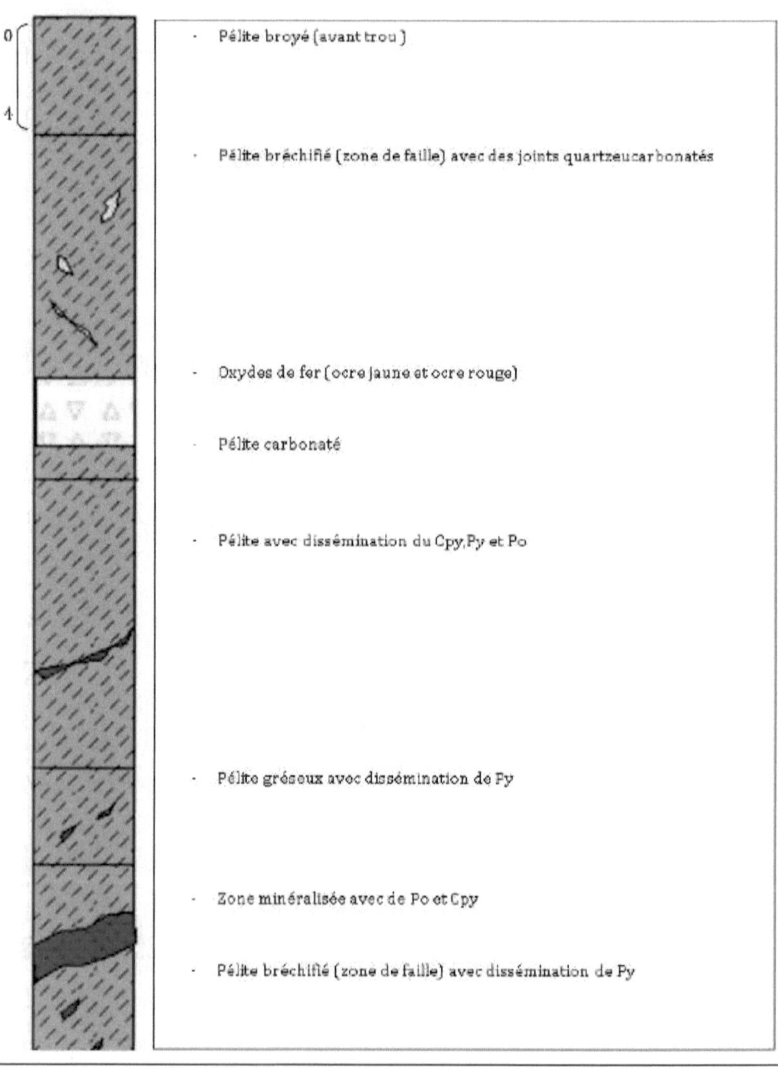

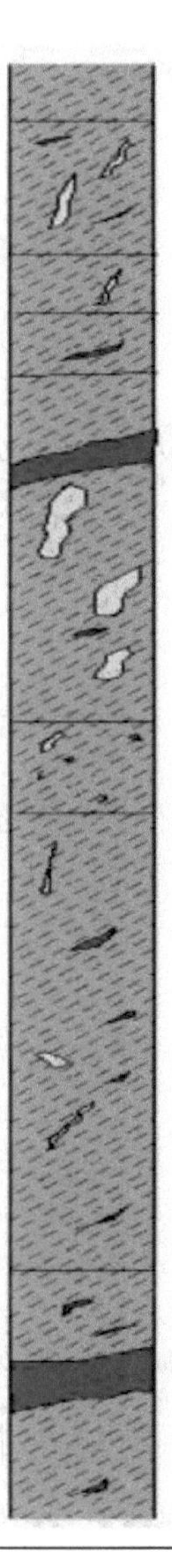

- Pélite avec des veinules de quartz; S1=45°

- Pélite avec une dissémination de Cpy, Py et Po ; S1=60°

- Zone minéralisée avec de Py,Po et Cpy

- Pélite bréchifié avec des veinules de quartz et une dissémination de Po (zone de faille)

- Pélite avec des veinules de quartz et une dissémination de Po

- Pélite bréchifié avec des veinules de quartz et une dissémination de Po (zone de faille)

- Pélite avec une dissémination de Cpy

- Zone minéralisée avec Po et Cpy

- Pélite avec une dissémination de Py

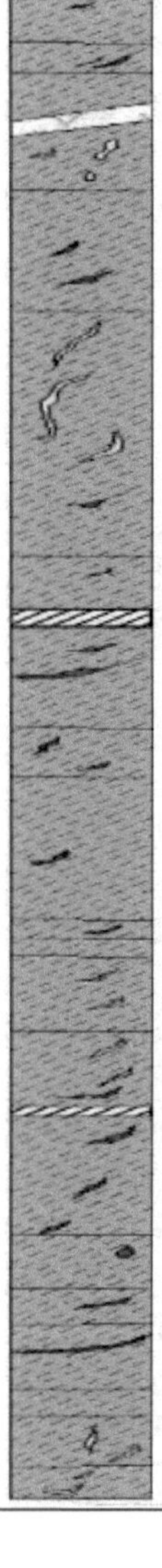

- Dyke avec des veines de quartz d'une taille centimétrique.

- Pélite avec dissémination de Po et Py plus des veinules de quartz

- Pélite bréchifié (zone de faille)

- Pélite avec dissémination de Py, Cpy et Po; S1−55°

- Pélite avec dissémination de Py et Po

- Pélite bréchifié avec dissémination de Po (zone de faille)

- Pélite avec des veinules de quartz ; S1−45°

- Pélite bréchifié (zone de faille)

- Pélite avec dissémination de Po et Py
- Pélite avec des plaquages de Po et Py dans les plans de S1; S1−55°

- Pélite sain ; S1−45°

- Pélite avec des joints de quartz ; S1−45°

- Pélite broyé (avant trou) ;

- Pélite bréchifié avec une teinte verdatre, des joints de quartz, S0 plissée et une dissémination de Sphalétire et Cpy ;

- Dyke avec dissémination de Py ;

- Pélite bréchifié et oxydé à l'intérieur avec des joints de quartz et dissémination de Py; S0 plissée (zone de faille) ;

- Dyke avec dissémination de Py et Cpy et des veines de quartz ;

- Pélite grisâtre avec plaquage de Po,Py et Cpy ;

- Dyke avec des vénules de quartz ;

- Pélite avec des joints de quartz et dissémination de Po et Py ;

- Pélite avec des veinules de quartz et plaquages de Po et Py dans les plans de schistosité ; S1=55° ;

- Zone minéralisée avec Po et Cpy ;

- Pélite avec dissémination de Po et Cpy ;

- Dyke avec dissémination de Py ;
- Pélite avec dissémination de Py et des veinules de quartz.

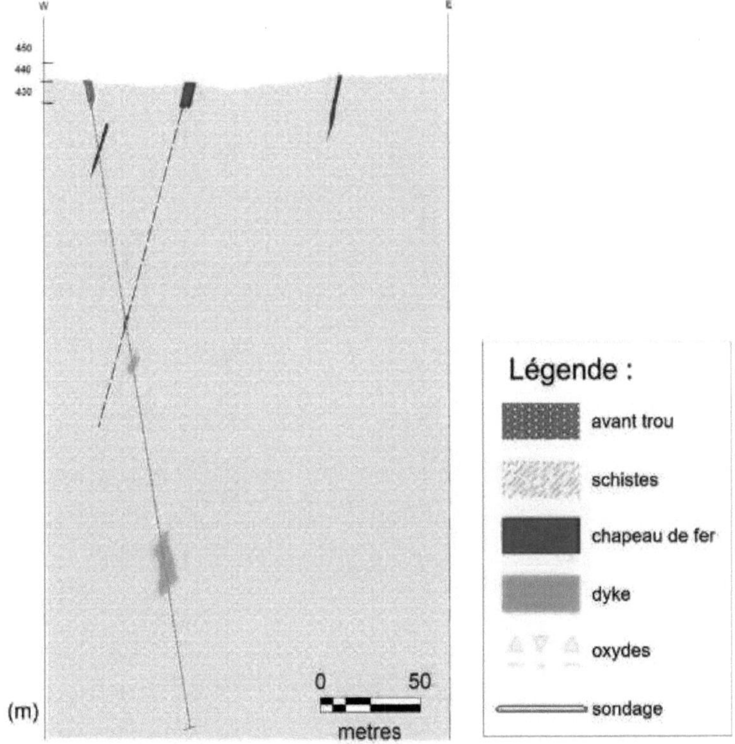

Figure 9: Geological section of borehole KTSC14

The section is oriented W-E and represents the KTSC14 borehole, which is inclined 80° to the East.

The pelite is crossed by basic dykes and steeply dipping mineralized lenses. The most distinguished lens in this hole is located at 109 m depth and has a diameter of 0.7 m in length.

We also note the presence of an oxidised zone of 2 m in size close to the surface at depth (just after the front hole).

Section of the KTSC16 survey:

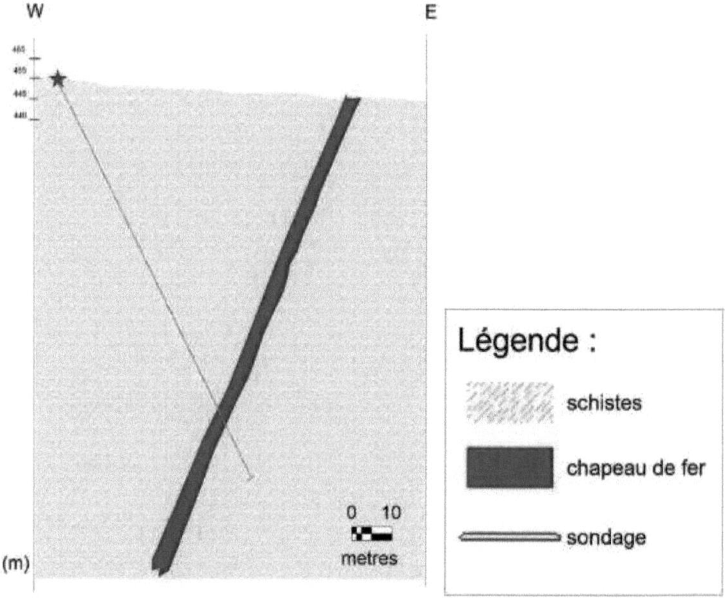

Figure 10: Geological section of borehole KTSC16.

The second section represents the KTSC 16 borehole, also oriented W-E, which is inclined 60° to the EAST.

Like the previous section, this section shows the dominance of the pelites which are crossed by the large iron cap inclined towards the West by about 60°. The hole hits the iron cap at a depth of 89.20m to 95m (5m of mineralization)

iii.   Mineralogical and gitological study of the mineralizations encountered:

Based on the study of core holes KTSC14, KTSC16, KTSC17. It can be concluded that the Kerkoz series consists of the wall to the roof of the iron cap.

<u>Basic training :</u>

Consists of the primary mineralization zone "Protore" where the dominant ore is pyrrhotite (iron sulphide), which has a metallic brown colour that changes immediately on contact with air or moisture.

The strength of the mineralization is between 10mm and 4m.

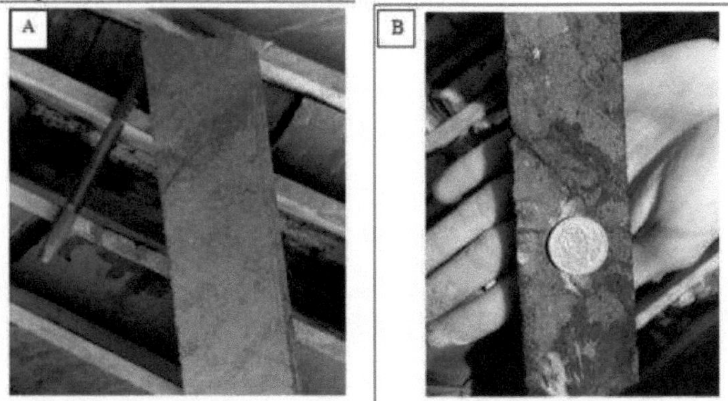

Photo 19: A and B showing the disappearance of mineralization in a core.

It is not an economic mineral but includes other minerals of varying economic interest such as chalcopyrite, pyrite, and sphalerite ....

The mineralization is crossed in places by quartz joints which sometimes contain pyrite or chalcopyrite.

Roof formation :

    - Oxidation zone :

Formed by iron oxides and hydroxides (hematite, magnetite) and yellow and red ochres which are due to the oxidation of the mineralization. The power of this zone is between 1m and 4m.

These ochres are somewhat hard due to the presence of amorphous silica.

Photo 20, oxidation zone in the borehole.

- Pelite :

Macrospectively, they are very schistose of varying strength.

Their fine granulometry combined with their schistosity, brings them closer to pelitic or clayey shales.

These pelites are characterised by a greyish colour, with the injection of millimetre to centimetre joints and veins of quartz, so that in places we can see some dissemination of sulphides in these pelites.

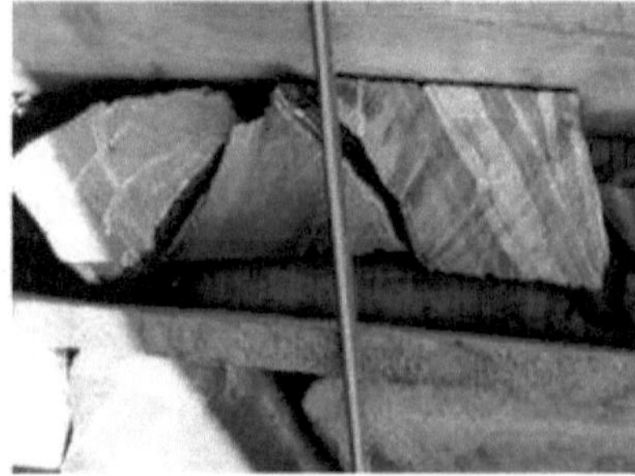

Photo 21: Schistosity planes (S1) in a core

Photo 22: Spread of pyrrhotite in pelites.

These pelites are intensely schistose with the S1 at an angle varying from 30° to 60° to the core axis, and there are zones of sulphide dissemination (pyrrhotite, chalcopyrite and pyrite) in the pelites, which tells us about the proximity of the mineralised zone.

38

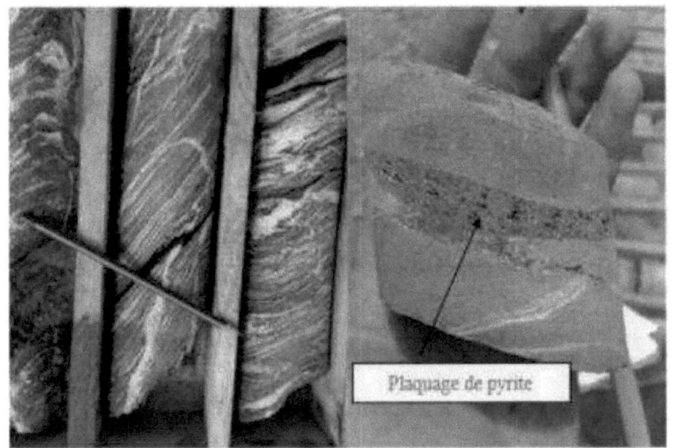

Photo 23: Mineralization plating in the chistosity planes and quartz joints in a core.

We note that this facies often presents alternating dark sandstone and light pelitic levels.

At the level of these pelites the stratification is generally at an angle of 45° to 50° with the core axis and sometimes it is found deformed and folded.

- Basic dykes :

It is an intrusive facies formed of either dolerite or gabbro, they have a fine texture where the minerals are not well visible with a blackish or greyish colour.

Photo 24: Dolerite dyke.

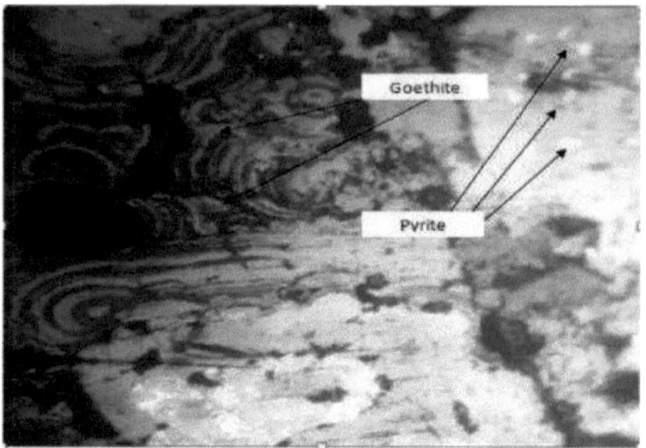

**Photo 25: Microscopic appearance of goethitemamel
and pyrite flies**

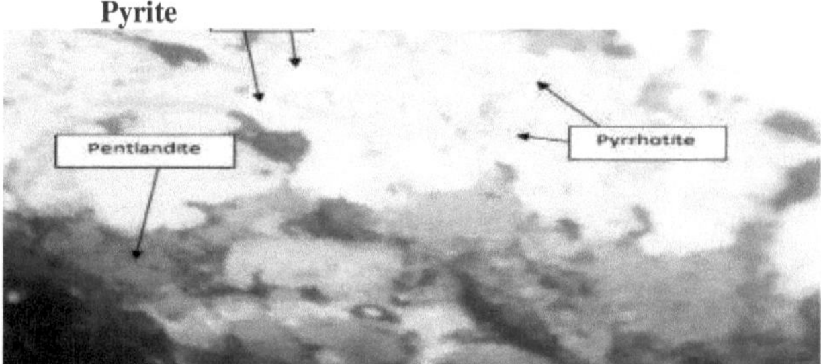

**Photo 26: microscopic appearance of pyrrhotite, pentlandite and
pyrite spots**

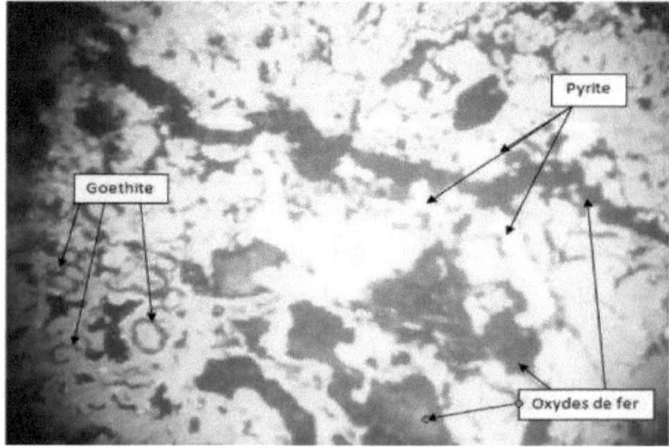

**Photo 27: Microscopic appearance of goethite, pyrite in the
form of flies and iron oxides**

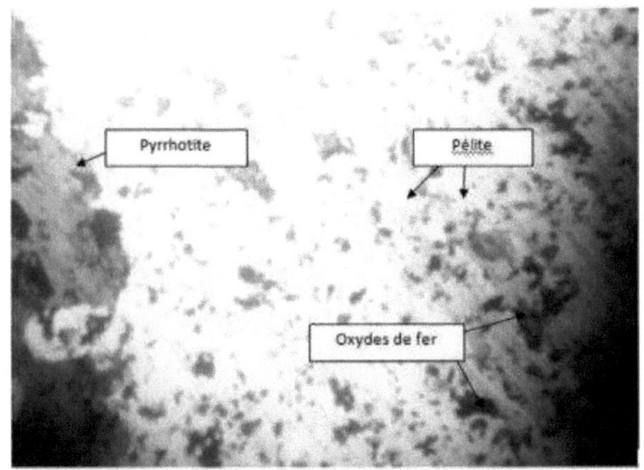

Photo 28: Microscopic appearance of pyrrhotite in the beach and a dissemination of pyrite in the pelites as well as iron oxides

6. Section made :

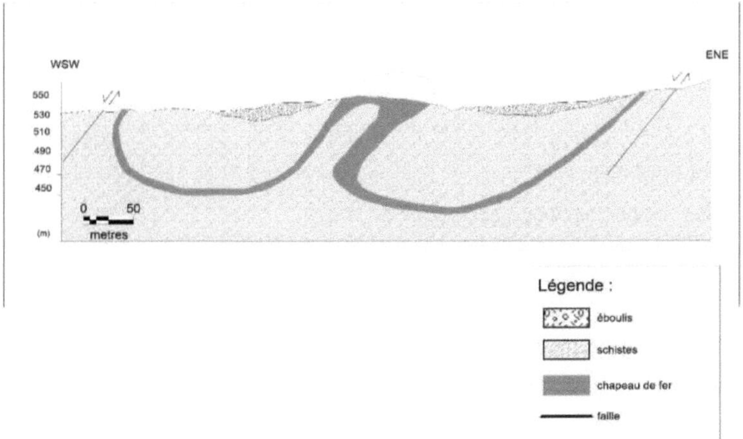

Figure 11: Geological section of the Kerkoz map

Section AB is oriented WSW-ENE where the formations are dated to the upper Venetian

It represents the folding of the iron cap at depth resulting from a flexible deformation (ductile cracking). This iron cap which is encased by the saghlef shales which are considered as the most dominant facies in our terrain has a dip parallel to that of the shales (between 45° and 60° towards the West), these shales are affected by reverse faults oriented N20 .

Then there is the overburden, which is in the form of scree of Quaternary age.

<u>Conclusion:</u>

According to the lithostratigraphic study of the terrain and the drill holes, the following mineralizations and facies are reported:

- The "Protore" mineralized zone, formed by pyrrhotite as the main mineral.
- Pelites which represent the bedrock of the study area.
- Basic intrusions that cross the pelites in the form of dykes.
- Strongly dipping quartz veins that are posterior to the host rock.
- The iron-rich oxidation zone which represents the iron cap roof

Roof formation: consists of the oxidation zone (iron oxide and hydroxide) and the pelites (host of the study area)

III.   The tectonics of the area :

The tectonics of our Kerkoz area is characterized by soft deformation (ductile shear, crenulation schistosity) and brittle deformation (sinister or dexter rifting, normal fault, fracture) and

Since the most widespread facies in our area is the Sarhlef shale, most of the deformations found are intra-schist.

1. Brittle deformation :

& Stalls :

The breaks occur in the contact zone between the carbonate rocks and the shales, they are either sinter or dextral, with directions between N80, N100.

<u>Photo 29: Dexterous detachment</u>

The change in the direction of the shales in the deformation zones shows that these breaks are later than the shales.

s **Loopholes :**

Loopholes without filling :

They are generally marked by the presence of indications of abnormal displacement: brecciation, fault spikes, abnormal discontinuity of iron caps and so on.

Photo 30; Brecciation showing the presence of a fault.

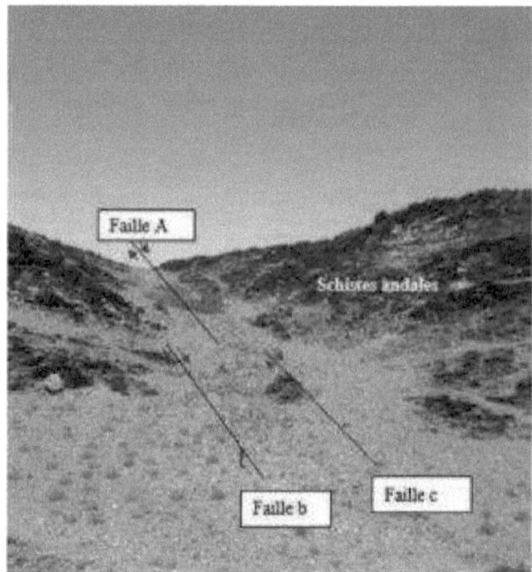

Photo 31: Three faults with the same direction

Rifts with filling :

They are in the form of veins or small hog's teeth veins or veins.
They are present everywhere in the mapped area.

Micro flaws :

In the core hole (KTSC17), we were able to observe the microtectonics that the Kerhoz zone has undergone, one of these deformations are the micro faults that dip parallel to the schistosity (45°) intersecting the mineralization disseminated in different zones.

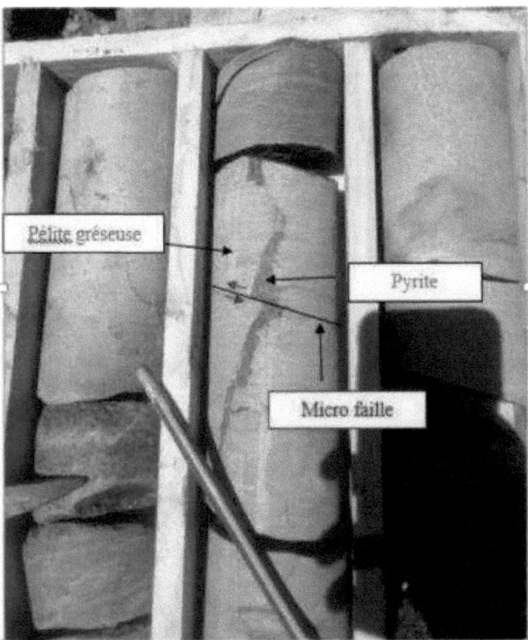

Photo 32: Micro fault in core sample KTSC17.

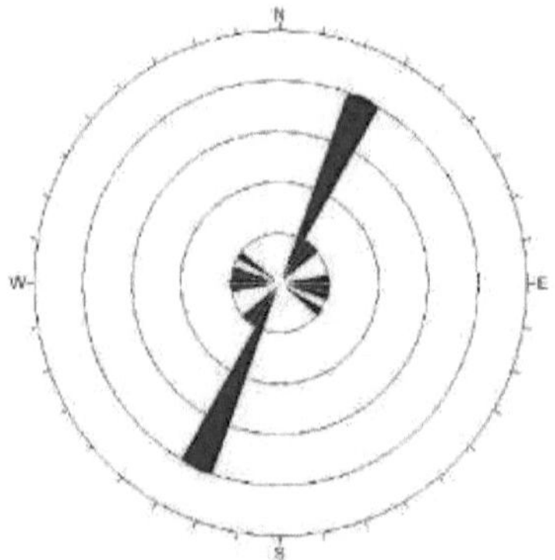

Figure 12: Rosette of faults.

From the rosette, we can see that we have 2 families of faults. A major one with a N30 direction and the other less widespread family with a N110 direction.

Fractures :

They are the most widespread in our area. They generally affect slab schists, dykes and also quartz veins, they are oriented N100 to 120 or N140 to N160.

Photo 33: Fractured quartz

Photo 34: Fractured shale

2. **Flexible deformation :**

46

Flow schistosity :

It is a flow of the rock in very fine sheets, with thicknesses ranging from microns to millimetres, and is accompanied by recrystallisations parallel to the planes of the sheets

Schistosity of crenulation :

Corresponds to zones of discontinuous shear fractures, almost parallel to each other, with millimetre to decimetre spacing, accompanied by folding that crosses the rock already affected by a first schistosity.

Photo 35: Photo showing crenulation schistosity.

Photo 36: Photo showing a folding of the shale (S2).

The cleavage planes are generally without recrystallisation. Minerals contemporaneous with the first schistosity are folded or sheared and in this case are sometimes arranged parallel to the cleavage plane (by simple mechanical reorientation).

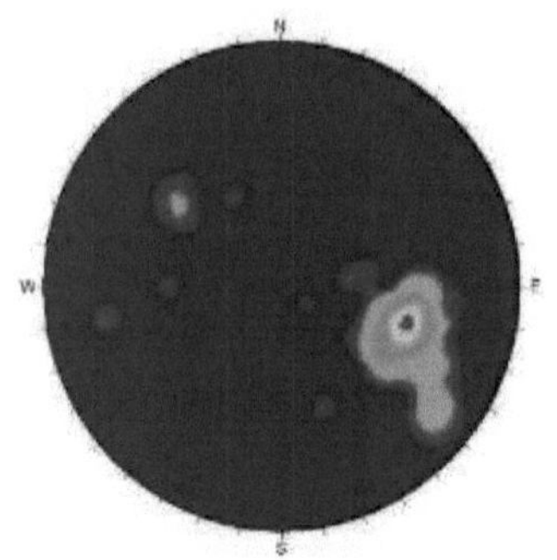

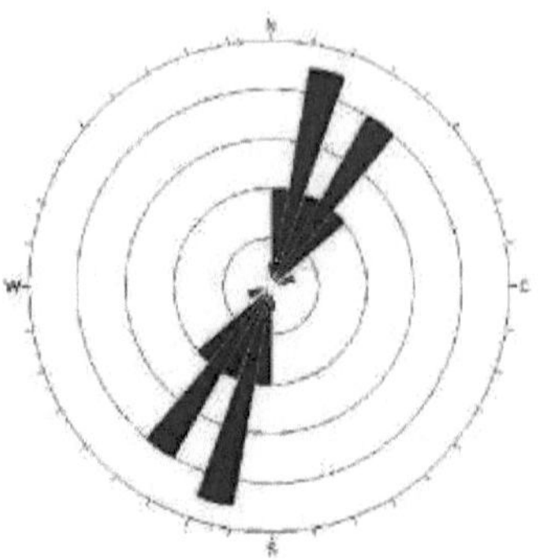

Figure 13: Schistosity rosette (inclination)

Figure 14: Schistosity rosette

The two schistose rosettes show that the main direction of the schistosity is N20 to N40 with an eastward dip.

3. **Ductile Deformation :**

From the observation of the iron caps and their morphologies on the ground and on the map, as well as the analysis of the core holes, we can conclude the mode of deformation that this area has undergone: a ductile deformation (shearing). This deformation manifests itself in the form of isoclinal folds inclined to the west by 60° and with a depth between 90m and 110m.

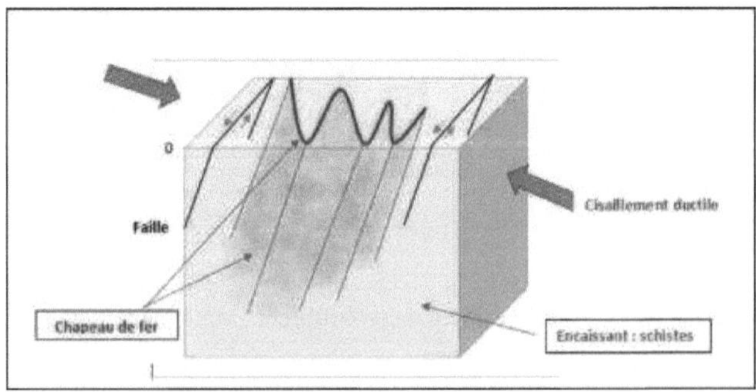

Figure 15: Block diagram showing folding of the iron cap

IV. Geochemistry of Kerkoz (Central Jbilets):

In this chapter, the distribution of metallic elements in the processed holes KTCS14 and KTSC 16 will be highlighted based on sampling of the mineralized impact.

Firstly, we measured the magnetic susceptibility of the mineralisation. We have tried to represent the distribution of the base metals, we would point out that the various analyses were carried out at the REMINEX laboratory.

Secondly, we will eventually study the links and correlations, which exist between the different metals based on a statistical study called PCA: Principal Component Analysis, using PCA software.

1. Distribution of base metals :

❖ **Results tables :**

KTSC16:

	Ag (%)	Cu (%)	Fe (%)	Pb (%)	Zn (%)	To (%)
KTSC 16 89.20 90.00	0,00038	0,0525	12,35	0,0995	0,0696	0,00002
KTSC 16 90.00 91.00	0,00072	0,54	41,84	0,0359	0,0581	0,000021
KTSC 16 91.00 92.00	0,00078	0,54	59,7	0,0251	0,23	0,000032
KTSC 16 92.00 92.70	0,00063	0,82	57,35	0,0213	0,0303	0,00002
KTSC 16 92.70 93.70	0,00081	0,43	49,54	0,0276	0,0318	0,000019
KTSC 16 93.70 94.20	0,00062	0,65	42,58	0,0228	0,0303	0,000014
KTSC 16 94.20 95.00	0,00034	0,1	25,68	0,0166	0,0473	0,000009
KTSC 16 94.20 95.00 **BIS**	0,00035	0,1	25,82	0,0158	0,0484	0,000012

Table 3: Distribution of metals in borehole KTSC16

	Ag (%)	Cu (%)	Fe (%)	Pb (%)	Zn (%)	To (%)
KTSC 14 108.5 109	0,00212	0,91	15,01	0,37	0,0655	0,000009
KTSC 14 109 109.7	0,00434	2,83	54,1	0,45	0,37	0,000009

<u>Table 3: Distribution of metals in borehole KTSC14</u>

2. **Variation in chemical elements at the borehole level :**

Variable	Comments	Obs. with missing data	Observations without missing data	Minimum	Maximum	Average	Standard deviation
Ag (%)	10	0	10	0,000	0,004	0,001	0,001
Cu (%)	10	0	10	0,053	2,830	0,697	0,807
Fe (%)	10	0	10	12,350	59,700	38,397	17,507
Pb (%)	10	0	10	0,016	0,450	0,108	0,162
Zn (%)	10	0	10	0,030	0,370	0,098	0,112
To (%)	10	0	10	0,000	0,000	0,0000165	0,000

<u>Table 4: Descriptive statistical table</u>

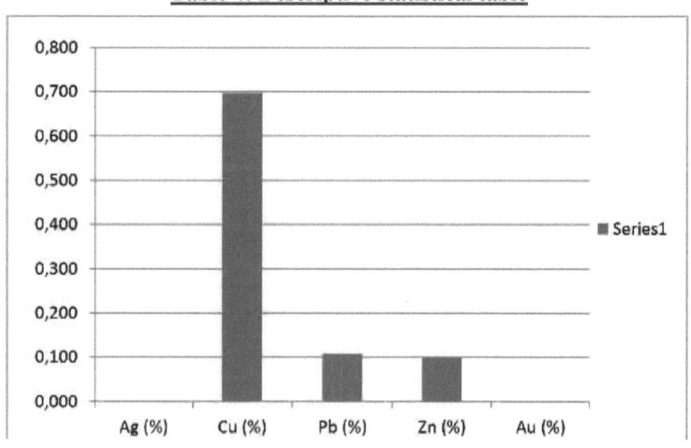

<u>Figure 16: Histogram of chemical element variations in boreholes</u>

✓ <u>KTSC 16 :</u>

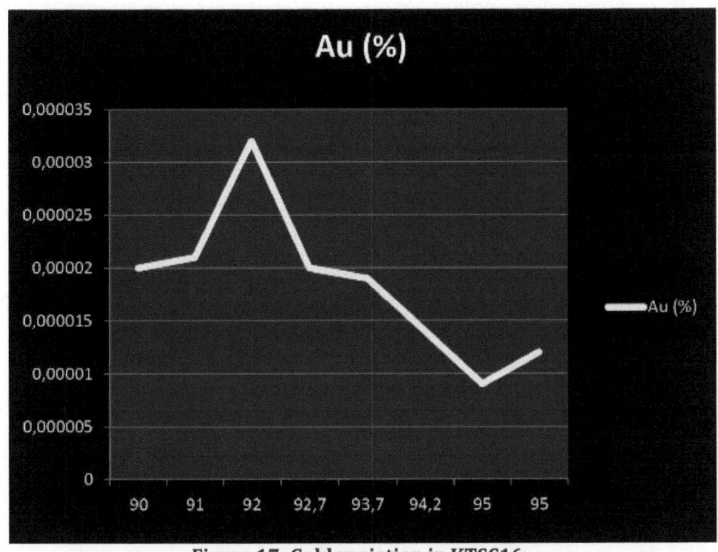

Figure 17: Gold variation in KTSC16

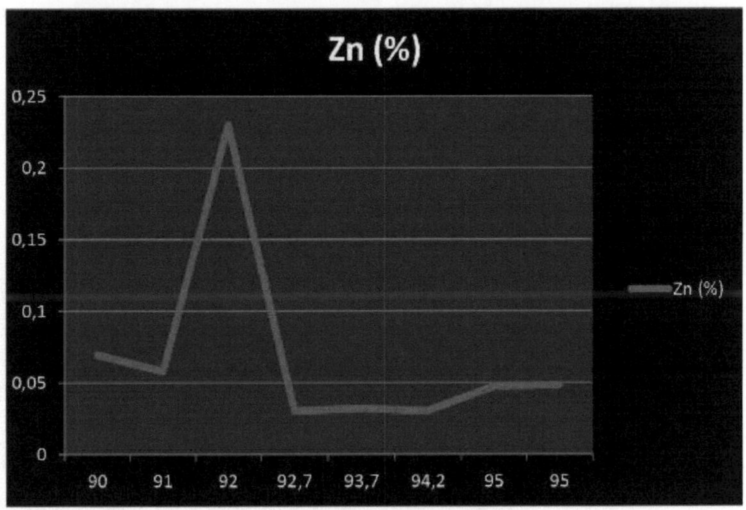

Figure 18: Zinc variation in KTSC16

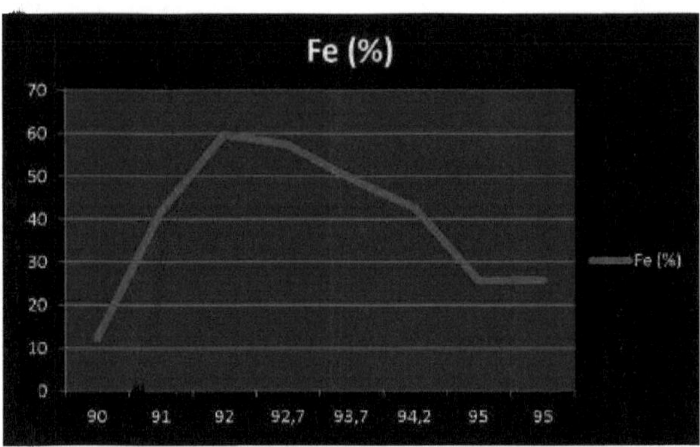

Figure 19: Iron variation in KTSC16

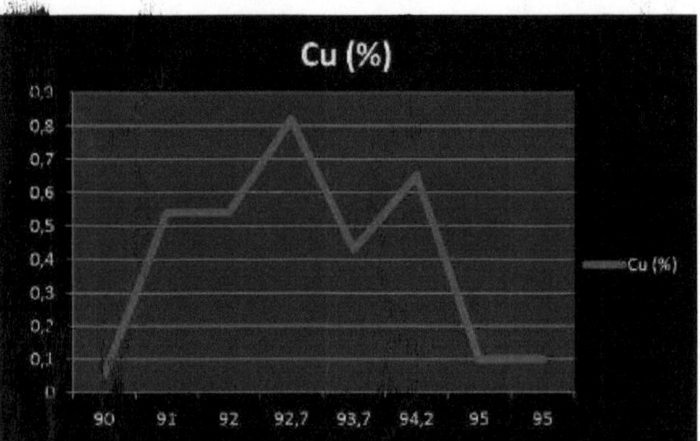

Figure 20: Variation of copper in KTSC16

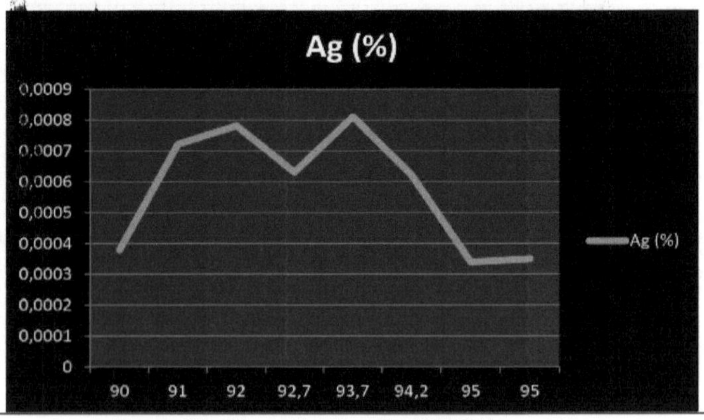

Figure 21: Change in silver in KTSC16

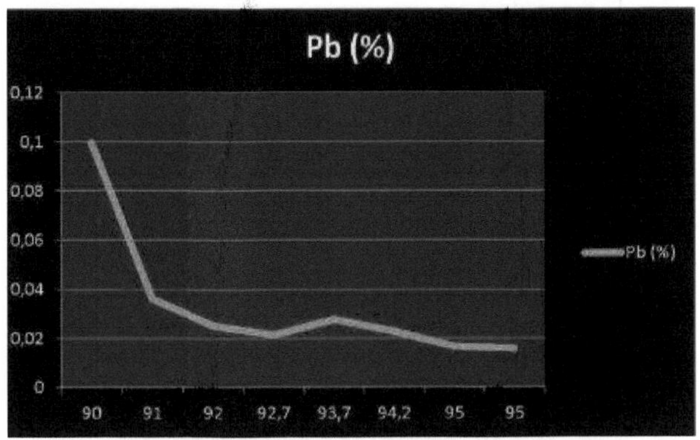

Figure 22: Lead variation in KTSC16

KTSC14 :

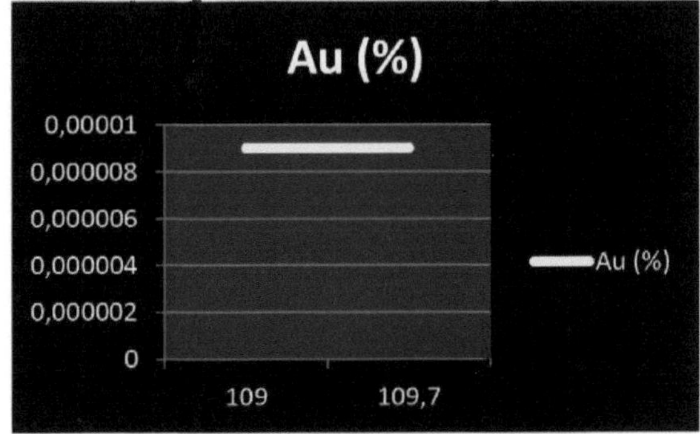

Figure 23: Gold variation in hole KTSC14

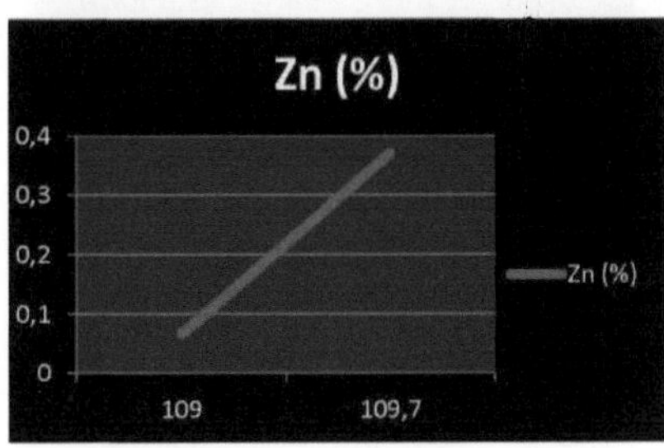

Figure 24: Zinc variation in borehole KTSC14

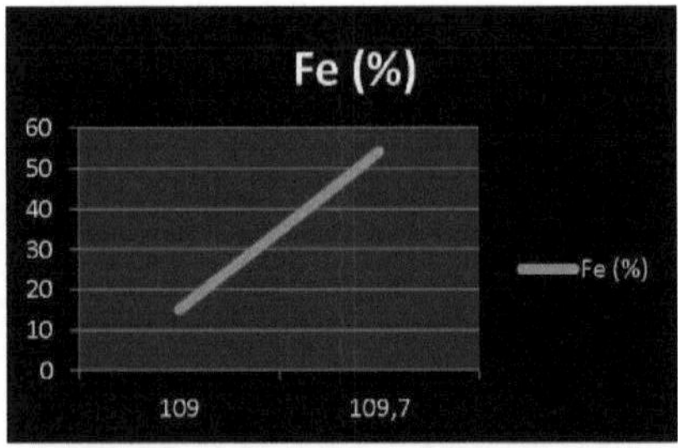

Figure 25: Iron variation in hole KTSC14

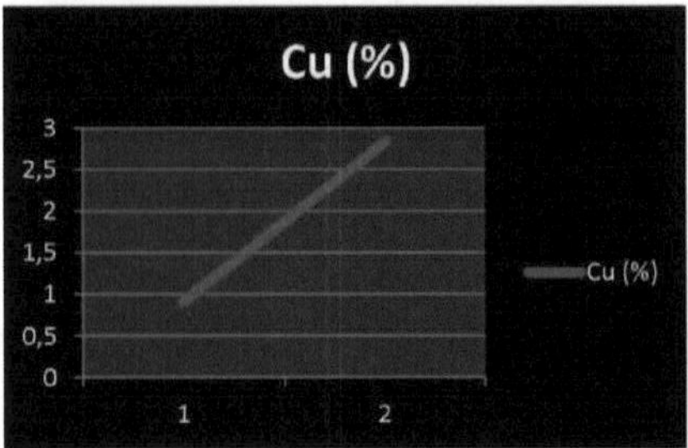

Figure 26; Copper variation in hole KTSC14

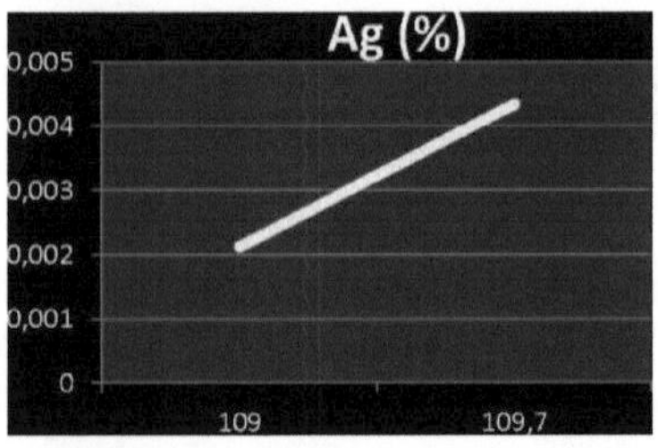

Figure 27: Silver variation in hole KTSC14

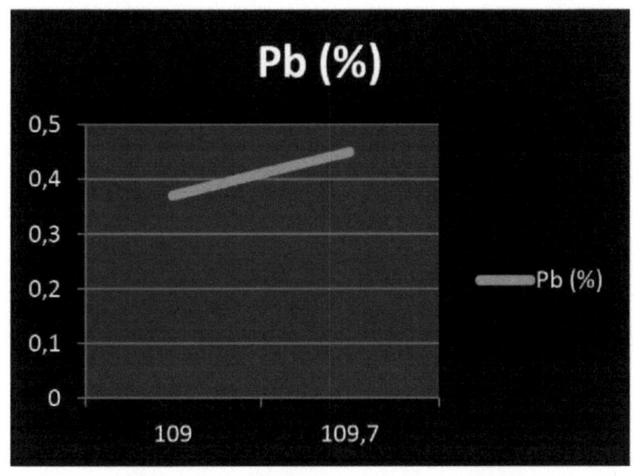

Figure 28: Lead variation in borehole KTSC14

3. Interpretation :

The distribution of the elements Cu, Zn, Pb, Au, Fe, Ag along the boreholes indicates that :

The iron content is very high along the boreholes;

> Pb levels are always lower than Zn levels;

> This variation in elemental content can be explained by the presence of different ore textures, or the presence of filled faults;

> the curves for Pb and Zn levels overlap.

4. Interpretation of susceptibility :

a) Susceptibility diagram :

KTSC14:

	Facies	K
E1	Pelite	0,4
E2	Yellow ochre	0,3
E3	Red ochre	0,1
E4	Carbonate pelite	0,2
E5	Pelite	0,7
E6	Pyrrhotite-Chalcopyrite	57,
E7	Pelite	0,4
E8	Pelite	0,7

E9	Pyrrhotite-Chalcopyrite	57,
E10	Pyrrhotite-pyrite	33,
E11	Pelite	0,7
E12	Pelite	0,4
E13	Pyrrhotite-Chalcopyrite	57,
E14	Pelite	0,4
E15	Dyke	0,3
E16	Pelite	0,7

Table 5: Table of susceptibility values in borehole KTSC14

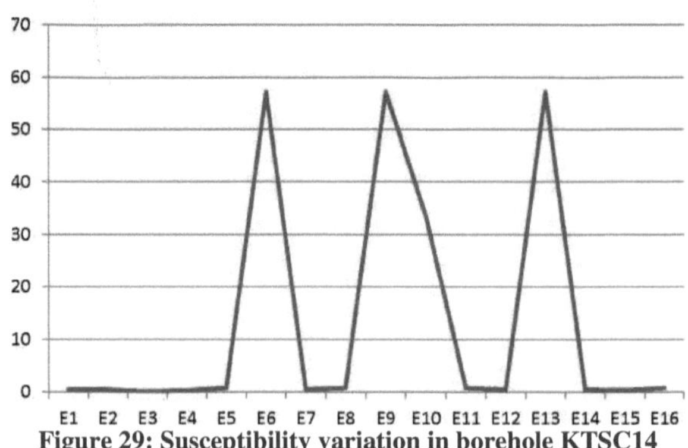

Figure 29: Susceptibility variation in borehole KTSC14

KTSC16:

	Facies	K
E1	Pelite	0,4
E2	Pelite	0,4
E3	Dyke	0,3
E4	Pelite	0,7
E5	Dyke	0,3
E6	Pelite	0,4
E7	Dyke	0,3
E8	Pelite	0,7
E9	Pyrrhotite-Chalcopyrite	57,
E10	Pelite	0,7
E11	Dyke	0,3
E12	Pelite	0,4

Table 6: Table of susceptibility values in borehole KTSC16

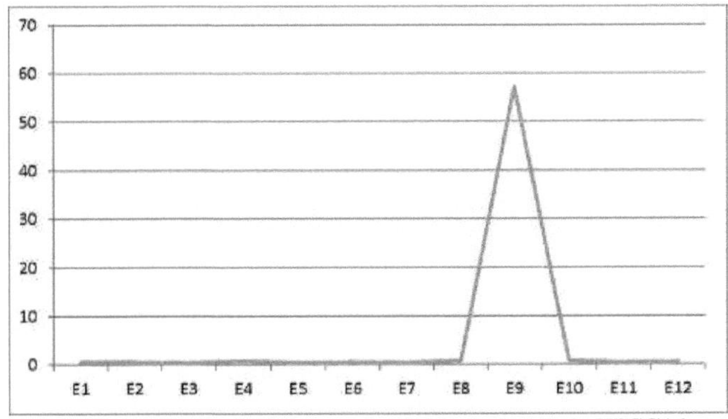

Figure 30: Variation in susceptibility in borehole KTSC16

KTSC17:

	Facies	K
E1	Pelite	0,4
E2	Pelite	0,4
E3	Dyke	0,3
E4	Pelite	0,7
E5	Dyke	0,3
E6	Pelite	1,0
E7	Dyke	0,3
E8	Pelite	0,4
E9	Dyke	0,3
E10	Pelite	1,0
E11	Dyke	0,3
E12	Pelite	1,3
E13	Dyke	0,3
E14	Pelite	0,7

Table 7: Susceptibility table of the high samples from the KTSC17 survey

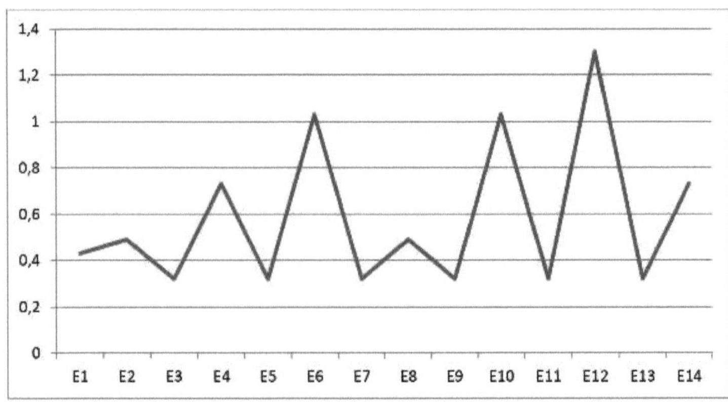

Figure 31: Susceptibility variation in borehole KTSC14

b) Interpretation :

From the susceptibility diagrams, it can be seen that the highest **K** values are measured at the pyrrhotite level where values range from 4 (the existence of enclaves in the host rock) to 50 (massive pyrrhotite).

The high susceptibility values in pyrrhotite are due to its iron-rich chemical composition (iron sulphide FeS).

We note the existence of low values (oxides and pelite) and high values in K which is expressed by the fluctuation of the curves.

<u>General conclusion :</u>

mapping, borehole studies, tectonics, and geochemistry of the central jbilet kerkouz zone have identified the following main features:

- The formations present in the kerkouz area are of upper Visean age, namely quartz veins, basic dykes, carbonates, iron caps, and the casing of all these previous formations: the Sarlef shales.

  The quartz structures (quartz veins) have a main direction from N0 to N30, these veins are generally located to the east of the studied area and correspond to the large N20 fault existing at this location.

- The kerkouz sector is characterized by a folding of iron caps in isoclinal folds resulting from the ductile deformation that this region underwent.

- The tectonics of the study area presents soft deformations (ductile deformation) which affected the studied iron cap in isoclinal folds, as well as brittle deformations (dry faults with a main direction N20 and others with a direction from N80 to N100, fractures, sintering and dextral faults...).

- Chemical analyses show high levels of iron and copper in the mineralized zones (pyrrhotite) but low levels of Pb, Zn, Ag and gold.

- Direct observation of the kerkoz zone in relation to the kettara zone, as well as the geological map and the petrographic and mineralogical study of this zone allowed us to construct the hypothesis that the mineralization of the kerkoz zone is an extension of the kettara zone:

Photo 37: Kettara-Kerkoz extension.

This hypothesis implied from the study of our area has led us to even propose the hypothesis that links all the existing prospects in the central jbilets including Kettara, Benslimane, Kerkoz, Koudiat Aicha. .etc.

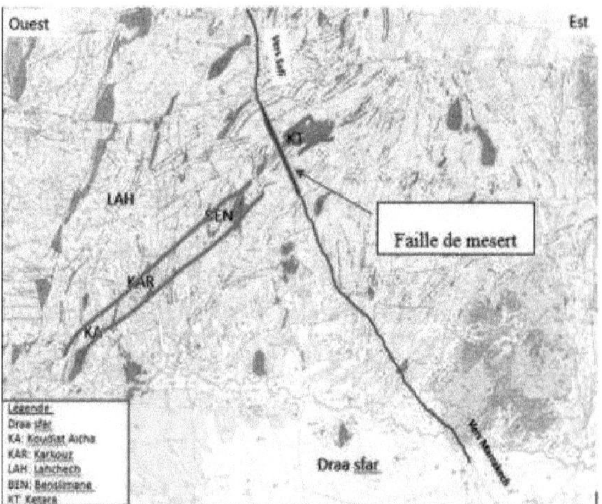

Figure 32: Geological map of the central jungle prospects

Bibliographic references

Bordonaro, M. (1983). Tectonics and petrography of the Kettara pyrrhotite district (Paleozoic Jebilet, Morocco). Thèse de 3èmes cycles, Université. Strasbourg, 132 pp.

Aarab, E. M. (1984). Mise en évidence du caractère co-génétique des roches magmatiques basiques et acides dans la série volcano-sédimentaire de Sarhlef (Jebilet, Maroc hercynien). Thèse 3ème cycle. Univ. Nancy.

Aarab, E. M. (1995). Genesis and differentiation of a tholeiitic magma in an extensive intracontinental domain, the example of the pre-orogenic Jebilet magmatism (Hercynian Morocco). Thesis of doct. d'Etat. Univ. Marrakech. 251 p

Belkabir et al., 2008. Geology and Wall-Rock Alteration at the Hercynien Draa Sfar Zn-Pb- Cu Deposit, Morocco, Ore Geology Reviews 33 (2008), pp. 280-306.

Essaifi, A. (1995). Relations entre magmatisme, déformation et altération hydrothermale, l'exemple des Jebilet centrales (hercynien, Maroc). Thèse de Docteur d'Etat Es-Sciences, Université. Cadi Ayyad University. Marrakech, 308 pp.

Hibti, M. (2001). Les amas sulfurés des Guemassa et des Jebilet (Meseta Sud-Occidentale, Maroc) : Témoins de l'hydrothermalisme précoce dans le bassin mésétien. Thesis of Doctor of State Es-Sciences. Université. Cadi Ayyad. Marrakech, 296 pp.

Huvelin, P. (1961). Sur l'âge Viséen Supérieur des Schistes de Kettara et de Djebel Sarhlef (Djebilet Centrales, Maroc). C. R. Sommaire Soc. Géol. France, 10: 290-291.

Huvelin, P. (1977). Etude géologique et gîtologique du massif hercynien des Jebilet (Maroc occidental). Notes et Mem. Serv. Geol. Maroc, 232 bis.

L. Ben Aissi, (2008) Contribution à l'étude gîtologique des amas sulfurés polymétalliques de Draa Sfar et de Koudiat Aïcha : comparaison avec les gisement de Ben Slimane et de Kettara (Jebilets centrales, Maroc hercynien) Thèse de Docteur Présentée à la Faculté des Sciences Semlalia Marrakech, Maroc ,313p

Lagarde, J. L and Choukroune, P. (1982). Ductile shear and syntectonic granitoids: the example of the Hercynian Jebilet massif (Morocco). Bull. Soc. Geol. France.

FSC
www.fsc.org
MIX
Papier aus verantwortungsvollen Quellen
Paper from responsible sources
FSC® C105338